1,000,000 Books
are available to read at

Forgotten Books

www.ForgottenBooks.com

Read online
Download PDF
Purchase in print

ISBN 978-0-265-72620-4
PIBN 10961623

This book is a reproduction of an important historical work. Forgotten Books uses state-of-the-art technology to digitally reconstruct the work, preserving the original format whilst repairing imperfections present in the aged copy. In rare cases, an imperfection in the original, such as a blemish or missing page, may be replicated in our edition. We do, however, repair the vast majority of imperfections successfully; any imperfections that remain are intentionally left to preserve the state of such historical works.

Forgotten Books is a registered trademark of FB &c Ltd.
Copyright © 2018 FB &c Ltd.
FB &c Ltd, Dalton House, 60 Windsor Avenue, London, SW19 2RR.
Company number 08720141. Registered in England and Wales.

For support please visit www.forgottenbooks.com

1 MONTH OF FREE READING

at

www.ForgottenBooks.com

By purchasing this book you are eligible for one month membership to ForgottenBooks.com, giving you unlimited access to our entire collection of over 1,000,000 titles via our web site and mobile apps.

To claim your free month visit:

www.forgottenbooks.com/free961623

* Offer is valid for 45 days from date of purchase. Terms and conditions apply.

English
Français
Deutsche
Italiano
Español
Português

www.forgottenbooks.com

Mythology Photography **Fiction**
Fishing Christianity **Art** Cooking
Essays Buddhism Freemasonry
Medicine **Biology** Music **Ancient Egypt** Evolution Carpentry Physics
Dance Geology **Mathematics** Fitness
Shakespeare **Folklore** Yoga Marketing
Confidence Immortality Biographies
Poetry **Psychology** Witchcraft
Electronics Chemistry History **Law**
Accounting **Philosophy** Anthropology
Alchemy Drama Quantum Mechanics
Atheism Sexual Health **Ancient History**
Entrepreneurship Languages Sport
Paleontology Needlework Islam
Metaphysics Investment Archaeology
Parenting Statistics Criminology
Motivational

THE
VICTORIAN NATURALIST

THE JOURNAL & MAGAZINE

OF THE

Field Naturalists' Club of Victori

VOL. II.

MAY 1885 TO APRIL 1886.

The Author of each Article is responsible for the facts and opinions he records.

South Melbourne:
MITCHELL & HENDERSON, PRINTERS, CLARENDON ST.
1886.

INDEX TO VICTORIAN NATURALIST.

VOL II.

	PAGE
Australia, Darwin on	20
Australian Birds, Oology of	126
Ballarat Field Club	43
Basalt-Vitrophyr, Notes on a	67
Birds, Notes on Habits of Native	90, 104, 140, 154
Botany, *Wormia Macdonaldi*	146
Chudleigh Caves, Trip to the	59
Coccidæ, Remarks on Victorian Gall-Making	99
Cockatoos and Magpies Habits of	154
Collector's Trip to North Queensland	109, 130, 139
Correspondence	63, 92
Anecdote of Duckling	64
Attempt to rear Cuckoo	63
Preserving Mixture	92
Crows, Notes on Habits of	90, 104, 140
Darwin on Australia	20
Errata	28, 40, 92
Eucalypts, Sanitary Properties of	84
Field Naturalists' Club	
Annual Conversazione	2
Exhibits	15
Lecturette, Extinct Animals	13
Insects	14
President's Address	3
Excursions	31, 33, 94
Brighton	31
Lal Lal	94
Lilydale	33
Exhibition of Wild Flowers	82
Proceedings, Monthly Meetings	1, 17, 29, 41, 53, 65, 81, 93, 105, 125, 137, 149
Flora, Additions to Queensland	74
Fungi of North Gippsland	106
Fungi, Victorian	76, 139
Geological Sketch of S. W. Victoria	70, 102, 114
New Guinea Plants, Notes on	19, 146
Notes	
Catalogue of Australian Hydroid Zoophytes	148
Foraminifera	28
Micro-Fungi	40
National Museum	40
Petrel Family	28
Select Extra-Tropical Plants	52, 147
Oology of Australian Birds	126
Orchids of Victoria	48, 142
Corysanthes	144
Microtis	142
Prasophyllum	48
Pterostylis	145
Plants, New Guinea	19, 146
Plants of Studley Park	24, 36
Queensland Flora, Additions to	74
Queensland North, Trip to	109, 130, 139
Studley Park, Plants of	24, 36
Tortoises, Note on Imbedded	103
Victoria, Geological Sketch of S. W.	70, 102, 114
Victorian Fungi	76, 106, 139
Wild Flowers, Exhibition of	82
Wilson's Promontory, Overland Trip to	43, 54, 87, 150

ERRATA.

Page 27, line 19—for "Flowers September to January" read "Flowers nearly all the year round."

Page 27, line 43—after "*Casuarina, Rumphius, Etym.* read "Supposed to allude to the leaves resembling the feathers of the Cassowary."

Page 69, line 29—insert "of the" after "felspar."

Page 69, last line—for "Weannie" read "volcanic."

Page 93, last line—after "lizard" add "*Grammatophora muricata.*"

Page 94, line 22—after "lizard" add "*Pygopus lepidopus.*"

Page 137, last line—after "and" insert "some of."

NOTICE TO BINDER.

No. 11, March 1886, is wrongly paged. Should read 137 to 148, instead of 125 to 136.

Vol. II. No. 1.   May 1885.

THE

Victorian Naturalist:

THE JOURNAL AND MAGAZINE

OF THE

Field Naturalists' Club of Victoria.

The Author of each article is responsible for the facts and opinions he records.

CONTENTS:

	PAGE
Proceedings of the Field Naturalists' Club of Victoria ...	1
Annual Conversazione.	2
President's Address.	3

PRICE—SIXPENCE

Emerald Hill:
J. C. MITCHELL, PRINTER, CLARENDON ST.
1885.

Field Naturalists' Club of Victoria.

OFFICERS 1884-85.

President:
Rev. J. J. HALLEY.

Vice-Presidents:
Mr. T. A. FORBES-LEITH | Mr. A. H. S. LUCAS.

Treasurer:
Mr. J. H. MATTHIAS.

Secretary:
Mr. F. G. A. BARNARD, Kew.

Assistant Secretary:
Mr. G. COGHILL.

Librarian:
Mr. C. FRENCH.

Committee:
Mr. D. BEST. | Mr. J. H. GATLIFF.
„ J. E DIXON. | „ C. A. TOPP.
Mr. H. WATTS.

Time of Ordinary Meetings—The Second Monday in each Month, at 8 p.m., Royal Society's Hall, Melbourne.

Subscription - - - Ten Shillings per annum.
For the "Victorian Naturalist," Six Shillings per annum, or Non-Members Seven Shillings. Post Free.

The Victorian Naturalist:

Vol. 2. No. 1.   MAY.   1885.

THE FIELD NATURALISTS' CLUB OF VICTORIA.

The monthly meeting of the Club was held at the Royal Society's Hall, on Monday evening, 13th April, 1885.

The president, Rev. J. J. Halley, occupied the chair, and about forty-five members and visitors were present.

The hon. librarian reported the receipt of the following additions to the Club's library :—" Science Record," No. 3 ; " Report of Ballarat School of Mines," 1883 ; "Proceedings of Ornithological Society of Vienna ;" "Supplement to Victorian Oology," Part 1, by A. J. Campbell.

The hon. sec. read the report of the sub-committee appointed to re-consider the list of Victorian birds proposed to be protected. It recommended that the following birds be struck out of the proposed list :—Hawks, bee-eaters, crow shrikes (except magpie, at present protected), finches, bower-birds, orioles, wattle-birds, leather-heads, and parrots (except swamp or ground parrakeet.) On the motion of Mr. Gregory, the consideration of the report was postponed till next meeting, pending replies from other societies, &c.

The hon. sec. read a short account of the excursion to Gipsy Village, Brighton, held on the previous Saturday, which had been well attended, and the members present were fairly successful in their finds. The rare orchids, *Eriochilus fimbriatus* and *Pterostylis apyhlla*, being obtained in bloom.

The following ladies and gentlemen were elected members of the Club:—Mrs. Beal, Mrs. C. W. Simson, Miss E. C. Simson, Messrs A. Campbell, J. P. Chirnside, R. A. Poole, and J. Russell, and Masters S. and D. Coghill, as junior members. Thirteen nominations were received for next meeting.

Messrs A. J. Campbell and J. E. Prince were elected to audit the accounts of the Club previous to the annual meeting.

Nominations for office-bearers for the year 1885-6 were then received, in each case the retiring office-bearers being the only persons nominated, except for hon. treasurer, for which Mr. Bage was nominated instead of Mr. Matthias. Three ladies and eight gentlemen were proposed as members of committee, being, Mrs. Dobson, Mrs. J. Simson, Miss Campbell, and Messrs Best, Gatliff, Hill, Le Soüef, Prince, Topp, Watts, and Wisewould. Mr. Best gave notice that he would move at the annual meeting that the number of members of committee be increased from five to eight.

Mr. J. E Prince, on behalf of Messrs Field and Son, of Birmingham, presented the Club with a valuable microscope for the use of the members, for which a hearty vote of thanks was tendered to him.

The hon. sec. announced that at the annual conversazione to take place on the 29th inst, lecturettes would be delivered by the Rev. A. W. Cresswell, M. A., on "The Extinct Animals of Australia;" and by himself on "Forms and Metamorphoses of Insects."

Papers read:—By Messrs Gregory and Lucas "Notes of an overland trip to Wilson's Promontory," Part I. Mr. Gregory read the descriptive part and Mr. Lucas the Natural History notes of the journey between the Trafalgar Railway Station and Mr. Mille's station at Yanakie, about two thirds of the distance travelled

The following were the principal exhibits of the evening:—By Mr. G. Coghill, five orchids in bloom, obtained on excursion to Gipsy Village, viz., *Eriochilus autumnalis*, *E. fimbriatus*, *Pterostylis aphylla*, *P. nana*, and *Prasophyllum Archeri*; by Mr. C. French, 260 species of Australian Coleoptera, family Buprestidæ, also orchids in bloom *Eriochilus fimbriatus* and *Pterostylis aphylla*; by Master C. French, carved gourd, from New Guinea; by Mr. G. R. Hill and Masters Hill, Victorian lepidoptera; by Mr. D. Le Soüef, living slow-worm; by Mr. T. A. Forbes-Leith, five British Birds, Rook, Common Gull, Black-headed Gull, Curlew, and Oyster-catchers, also pair of Opossum mice ; and by Mr. F. Reader, plants from Studley Park, (orders *Apocyneæ*, *Solanaceæ*).

After the usual *conversazione* the meeting terminated.

FIELD NATURALISTS' CLUB OF VICTORIA.

Annual Conversazione.

The Fifth Annual Conversazione of the Club was held at the Royal Society's Hall on Wednesday evening, 29th April, 1885, when there was a very large attendance of the members and their friends, it being estimated that over 350 ladies and gentlemen were present.

On their arrival the visitors rambled through the lower rooms of the building, which contained a very fine display of objects of natural history both living and dead The tables were arranged on a much better plan than last year, and allowed greater facilities for studying the many excellent and beautiful exhibits, which will be fully detailed further on. Prominent among these may be noticed the magnificent collection of Australian parrots, shown by Mr. T. A. Forbes-Leith, the case containing representatives of 65 species; the many rare beetles, butterflies, and moths exhibited by Mr. C. French; the fine collections of shells by Messrs Gatliff and Worcester; the lepidoptera of Mr. Kershaw; the Australian coleoptera of Mr. Best; the Victorian sponges by Mr. Lucas; the live snakes by Mr. D. Le Souëf; the rare plants by Baron von Mueller; and the growing Victorian ferns by Mr. F. G. A. Barnard.

After a pleasant half-hour among the birds, insects, etc., the visitors assembled in the upper hall to hear the Rev. J. J. Halley deliver the presidential address, which was as follows:—

Ladies and Gentlemen, Members of the Field Naturalist Club of Victoria,

In the address, which custom assigns to the President of a society like ours, at its annual gathering an opportunity is given for a deliverance on any great subject that may have agitated intellectual society, or work done may be reviewed, or suggestions for future operations may be advanced. But before I attempt to do my poor part in any one of these directions it must be mine to thank my fellow-members for the very high honour they have conferred upon me in unanimously and cordially electing me to be their President —an honour alike unsought and unwished for. Ladies and gentlemen, while I thank you for this honour, I think that you have made a mistake. Your President should be one who, in the arena of science, has won his knightly spurs like my learned predecessors, Professor M'Coy and Dr. Dobson, rather than one who pretends to be but an esquire, achieving no conquests for himself, but merely bearing arms after nobler combatants. I may, at any rate, congratulate our Club on the pleasant and prosperous year that now draws to a close. Our meetings have been always interesting and instructive, and sometimes specially so. Rare and costly specimens in all departments of natural history have graced our exhibitions. Papers not unworthy of more ambitious societies have been read, honest work in the field has been done, and we number in our guild 160 ladies and gentlemen.

Nor need we fail to congratulate ourselves that, of the learned societies of Victoria, we have been the first to recognise that there are priestesses worshipping in the temple of Nature. Other societies have invited ladies to grace and add sweetness and lustre to annual gatherings, or occasionally, in a kind of superior patronising way, have arranged special evenings when more serious work was dispensed with, and curious or pretty things were shown or said, fitted to what was evidently deemed the taste of weaker intellects, but not only thus we meet on gala days in festive dress, but to share with us in honourable toil, side by side to delve in intellectual mines—to make common explorations into undiscovered lands of science—to strive to make nature give up her secrets, recognising in the fullest sense a common inheritance and a common right. The roll of our membership bears the names of 20 sisters of science. With the higher education of women an accomplished fact, with a girls' college in this city distancing in matriculation honours all the boys' grammar schools and colleges, I am sure of this, that whether we men will or will not, sooner or later we shall have to open, without distinction of sex, the doors of all our intellectual and scientific societies, and I trust that it will be our privilege, before many years have passed, to listen to this annual address delivered by one of the sisterhood of our guild.

It is evident that this action of ours looks far beyond the mere admission of ladies to our meetings, and it is for this that I dwell upon it, for we cannot but recognise that it must play no unimportant part in what may be called " the domestication of science." We may be thankful that at last, however inadequately, natural science forms a part of the curriculum of most of our higher schools. The more common phenomena of nature are, at any rate, investigated and explained, and principles are more or less discussed. Collections of fauna and flora are common in our homes. Microscopes are found in nearly all studies. The happy home is certainly the intelligent home—the home where each member is able to add something to the common stock of thought and knowledge, and, as has been said, " where the family does not consist of an ill-assorted aggregation of babies, great and small, dependent for their amusement upon some rattle of frivolity, or the chance of a stranger tickling them with a fashionable straw." The increase of our intelligent and happy homes has been brought about by the increase of our intelligent mothers and sisters. Cynics will, doubtless, say that the majority of our young men care far more for sport than science, for cricket than for conchology, for football than for floriculture, for rifles than for reflection ; and that mothers must bring up girls to suit the taste of the market, whatever it may be—if the demand be for frivolity, frivolity must be produced ; if for stupidity, stupidity must be forthcoming. We may hope that the cynic's sneer is fast losing its sting—that the demand for frivolity, ignorance, or stupidity is getting to be at a discount ; and to the women of our own day, members of our Club or not, we will quote the words of that great master of science, Sir Humphrey Davy, in an appeal to women made seventy-four years ago : " Let them make it disgraceful for men to be ignorant, and ignorance will perish ; and that part of their empire founded upon mental improvement will be strengthened and exalted by time, will be untouched by age, will be immortal in its youth." Of all schools of knowledge after those of music, painting, and sculpture, natural science is the best adapted for domestication. Some departments of intellectual investigation seem to adapt themselves more to the study than the parlour—to invite their devotees to solitude rather than to company ; but the pleasure of a discovery in the world of nature is more than doubled by being shared; and the pathway to its mountain heights is made easy when travelled in company. In this colony of ours, with all its exuberance of youth, with all its free, wild life, with all its deification of manly sports, the domestication of science will help to teach

> That life is not an idle ore,
> But iron dug from central gloom,
> And heated hot with burning fears,
> And dipt in baths of burning tears,
> And batter'd with the shocks of doom.

We have fallen on utilitarian days. Societies have to show that they have a right to existence ; a *raison d'être* is demanded from all. Our answer to the challenge thrown down then is, that we exist for the purpose of popularising science—of fostering a love for nature—not by the mere study of what other men have seen or the examination of theories propounded by the giants of our race—but by examination for ourselves in the field. Not that the study of books is to be neglected—none of us can afford to do that—but to use our book knowledge as a guide to our field investigation, and by actual observation for ourselves to verify or otherwise what books have taught us. In this learning we must be content to be patient, reverant, childlike, not too hasty, from imperfect data, to jump to conclusions—nor yet, when we get undoubted facts, too conservative to give up any pre-conceived opinions or theories. Starting from our books, going into the field, observing, arranging, theorising, we shall need to understand how, on the one hand, to avoid the Scylla of wild speculation, and on the other hand the Charybdis of mere antiquated and worn out belief. The more we learn the more modest we shall doubtless become ; it is the tyros, not the veterans, who are sure about everything—the many times that we have to modify our opinions will teach as modesty of expression. But if we are true students of nature we shall never tire of listening to her teachings, for she will lead us into a veritable fairyland, and she will tell us wondrous tales. To her children nature is as Longfellow makes her in his poem on the birthday of Agassiz—an old nurse—and she sings to her children thus—

> Come, wander with me, she said,
> Into regions yet untrod,
> And read what is still unread
> In the manuscript of God.
>
> And he wandered away, away,
> With Nature, the dear old nurse,
> Who sang to him night and day
> The rhymes of the universe.
>
> And whenever the way seemed long,
> Or his heart began to fail,
> She would sing a more wonderful song,
> Or tell a more wonderful tale.

The study of nature is no longer a hidden mystery, to be unveiled only to a few initiated ones. The days when the goddess was carefully hidden from the gaze of the common people, guarded by priests, jealous lest any save themselves should behold the Deity, have passed away. Isis has been unveiled, and all who will may, by living study, enter into the most secret recesses of the fane. Again, then, we affirm the aim of our Club is the popularisation and domestication of science.

I ask, next, what are the facilities afforded for the study of natural science in this colony of ours? The wisdom of the founders of institutions in this young land has been shown by the liberality with which provision has been made for the study of art and science. Our public library, our picture gallery, our botanical gardens, zoological gardens, and museums are the pride of our city, and a wonder to those who remember that not a century has passed since one was "first to sail into a silent sea," and barely fifty years since white men made a home where our stately city now stands. That these liberal provisions were made none too soon is evidenced by the fact that there is hardly a literary or a scientific society of the old land that does not find its counterpart here, and it is indeed to be hoped that Australia's children may not only hold their own in the cricket field, not only fight side by side on Africa's sands with England's sturdiest, fired by a noble, if, perchance, a somewhat wild ambition, but also win their laurels in the arena of literature, science, and art.

As I have already intimated, the first need of a student is books —books to guide him in the way he wants to travel. Of manuals dealing generally with scientific subjects or treating of great principles our private and public libraries are well supplied. Botanical and zoological text books are not difficult to obtain, but what we do need is books dealing specifically with the various departments of fauna and flora as they are found in this colony. This need was pointed out by my learned predecessor in this chair, Dr. Dobson, who last year pleaded for a "Dichotomous Key to the Plants of the Colony," and so well did Dr. Dobson plead, and so wisely did he act, that the Government Botanist, the Baron von Müeller, readily undertook the preparation of such a key, and has, during the past year, given to it much attention, and hopes ere long to have it ready for publication. I understand that this key is to be made as useful as possible in that it will be illustrated. Those of our members who make botany their study are to be congratulated on the prospect of so soon having their labours lightened.

But what Baron von Müeller is doing for plants is much needed in all departments. Our students find it very difficult to learn what objects have or what have not been described. A strange shell, or mollusc, or zoophyte is found, and there is nothing to tell if it be new to science or not; often even its generic place is hard to discover. What we need are monographs or catalogues. So far as one family is concerned, this want has been, during the past year, ably supplied by the publication of a catalogue of "Australian Hydroid Zoophytes." We are indebted for this immense help in the study of forms familiar on all our sea coasts to Mr. W. M. Bale, Secretary of the Microscopical Society of Victoria, and a member of our Club. Mr. Bale has described and figured nearly 200 forms, diligently searching previous records, and so presenting to us a catalogue made up to date. The illustrations, so carefully and accurately lined, will enable, in connection with the descriptions, the observer to identify and name any of the forms that may come under his notice. An introduction of 40 pages gives a sketch of the most important features of the structure and life history of the Hydroid Zoophytes. We cannot, however, while congratulating the author on the useful and important work he has produced, but express our regret that the book is headed "Australian Museum," rather than that of the National Museum of Victoria, and the imprint of Thomas Richards, Government Printer of Sydney, rather than that of John Ferres, Government Printer of Melbourne. It is hardly creditable that the bringing out of an important work on Natural History by a Victorian student should have been undertaken by the Government of another colony.

Nor must I pass without notice a catalogue of the eggs of Victorian birds, by Mr. Campbell, together with a supplement. While Gould left little to do, so far as the birds of Australia are concerned, he was not able, from the comparatively short time he was in the colonies, to tell us much of the oology of our Aves. Now that students are directing their attention to life histories, a knowledge of embryology has become a matter of great importance, and, in connection with embryology, oology is likely to assist in the elucidation of many mysteries. During the year Prof. M'Coy has been enabled to publish the ninth decade of the "Natural History of Victoria." The first of these decades bears date the 24th June, 1878, so that at the present rate we get the description on an average of fifteen forms a year. Our mathematical friends will be able to calculate at what distant period the memoirs of our museum will be complete, and geologists may perchance dream as to what will be the geological state of our earth when the last plate shall be issued, and how many of the present living forms will then more fitly find a place in a paleontological record. As to the prodromus of the paleontology of Victoria, the last decade bears date 1st of September, 1881. Since that date many collectors of organic remains have been anxiously and patiently waiting for their description. It is much to be regretted that the able and learned professor, whose accurate knowledge none can doubt, is unable, from his numerous and important engagements, more frequently to issue these helpful and beautifully illustrated papers, for it can hardly be that the Government of so wealthy a colony grudges the sum required for their production. True students will, however, battle on with or without aid, and doubtless difficulties will only incite to noble effort.

In this connection I cannot but mention the "Forest Flora of South Australia," by Mr. J. E. Brown, Conservator of Forests in that colony. The size and beauty of the plates will charm all the lovers of our native woodlands. Nor must we forget our own modest manual of the Club's proceedings, "The Victorian Naturalist," of which our first volume has been published, and in which will be found many papers of interest—a baby yet among such like productions, but promising to grow bigger and stronger as members of the Club enrich its pages by their observations, and increase its circulation by their efforts.

Passing from the literature of our subject, we come to collections of specimens alive or dead. The student of animal life can spend many pleasant hours in the zoological collection at the Royal Park, and if he has the good fortune to secure Mr. Le Souef as his guide, philosopher, and friend, his pleasure will be doubled. Whatever blame may possibly, and only possibly, rest on the shoulders of the Acclimatisation Society of Victoria in respect to some of their introductions into the colony, nothing but praise can be awarded to them for the care and enterprise that has been shown in the collection and management of their gardens—gardens that will surely induce a love of natural history in the minds of young

Victorians, but which also prove of great value to the student who wants to study the habits of beasts, birds, and reptiles. However skilful a taxidermist may be, he can never give to his skins the subtle and mysterious quiver of life, so that the student who wants to understand life and its history seeks to learn from living objects, and the gardens of the Acclimatisation Society meet a felt need.

The wealth of our city in this direction has been added to by the opening of the aquarium in the Exhibition Building. It is true that at present but few species have been secured, and that whole classes of marine fauna, such as the *Actinozoa*, that make some of the tanks at Brighton, England, gay as tulip-beds, are conspicuous by their absence, yet enough has been done, and well done, to show what the possibilities are, and doubtless the management, which has made so good a beginning, will not rest till the icthyologist finds, not only something to amuse, but opportunity for grave study. The names of the inhabitants are well and conveniently set out on the tanks ; but, for the sake of the many who have no knowledge at all of fish, it would be well, in cases where more than one species are in the same tank, to give a description, brief but clear, so as clearly to indicate which is which. It is amusing to listen for a little while at one of such tanks, and note the strange guesses made, and the stranger reasons given for the belief entertained.

We have four museums, all of which demand attention, and render aid to the student of natural history, and should enable the collector to name most of his finds, and so to put him in the way of studying correctly life histories. In this way home collections will be more than pretty toys, and the aim of our club and the subject of our paper will begin to be realised—the domestication of natural science.

Of our National Museum, for its large collection and the admirable way in which the taxidermist has arranged many of the groups of birds and animals, we have a just right to be proud. Having visited many of the natural history museums both at home and on the Continent, our own, I can safely say, in many respects, contrasts most favourably with these, in some, carries off the palm for excellency. But there is here yet much to be desired, and a deputation from our Club waited on the trustees of the Library and Museums for the purpose of pointing out to them some requirements. I may mention them here :—

1st. The first great need is more room. Treasures are there, but they cannot be found. Entomolgy is a favourite department of science in this colony, and the collections of insects are numerous. Some enthusiast, proud of his gatherings, makes his way to the National Museum to identify his species. He looks, and often looks in vain. A few cases—many obsolete names—and yet the museum is rich in such gatherings, only they are stowed away. By the courtesy, indeed, of Mr. Kershaw they may be seen, but the many, I am afraid, do not know the amiability of our fellow-member. The same complaint may be made as to oology. It is believed there is a good oological collection somewhere, but where that somewhere is no ordinary visitor can find out. The *Sauropsida* of Australia form an interesting study, and our museum ought to contain a fairly complete series. But here, too, we look in vain. How beautifully reptilia may be preserved and mounted, and made, instead of repulsive, almost fit for drawing-room ornaments, visitors to the newly established museum at Adelaide can testify. Without further illustration, what is sorely needed at our National Museum is room—room for the arrangement especially of the fauna of Australia—that our museum may not be simply a lounging place for the nursemaids of Carlton, or a show for passing visitors, but a place where our students of nature can find the real helps they need. It has also been pointed out that in many cases the nomenclature is antiquated, and in some cases inconsistent.

2nd. We ask that the overworked learned professor at the head of the museum should have given to him a staff of scientific assistants. Our idea is that, under Professor M'Coy, there should be a number of gentlemen, each one with the care of a department. We could not expect such a list of world-known men as form the staff of the British Museum, nor would it be necessary. There are plenty of young scientists who would be willing to be working heads, and who, under the direction of the professor, would be able to collect, classify, arrange, name, and, in addition, be able—not to waste time by chatting with idle *dilettanti* or answering foolish questions—but to put enquirers into the way of finding solutions to their seekings. No one man can do or ought to try to do everything. Our colony is rich enough, and the students of nature are many enough, to warrant such an arrangement. It would be ungenerous to blame an overworked man, yet it is intolerable that specimens should, in the last two or three years (to my own knowledge), have over and over again been sent to Europe for identification ; of such specimens not a few proved new to science. We ought to be able in this colony, at any rate, to classify and name our own natural productions. With increase of room and increase of men the other needs mentioned by our deputation to the museum trustees could easily be met.

Our second museum is the technological one at the Public Library, of which little need be said ; it is very useful, so far as it goes. The ethnographical department bids fair to be of much use to those who desire to study Polynesian races—a department that should be much increased by the acquisition of specimens of the dress, arms, implements, and works of art of the aboriginal people of Australia and Austral Polynesia. Many races seem doomed to extinction ; before it is too late let us preserve all we can that may teach us and those who come after us what sort of men they were.

The last born of our natural collections is the Economic Museum at the Exhibition Building. The danger of this new undertaking is that it should overlap on the one hand the Nationl, and, on the other, the Technological Museums. To a certain extent this has already been the case. Conchological and paleontological collections should certainly find no place there, that is if our museums are to be helps to study and not mere show places. To be valuable, collections should be complete. The authorities of our various institutions should work together in harmony and with common purpose. Public money should certainly not be spent in gathering a few specimens at the Exhibition Building of shells, or fossils, or the like, and even presentations of such should be handed in to the National Museum. A student cannot afford time, if he needs to compare specimen with specimen, to run between the Exhibition and University Buildings. But an Economic Museum in itself is another and much needed help, not only to the scientist, but to those who are called the practical men of the community—manufacturers, agriculturists, horticulturists, all need such an aid. Specimens of products, with their economic uses ; complete sets of insects noxious to plants, such as have been prepared and placed there by our fellow-member, Mr. French ; the admirable series of woods by Baron von Mueller; complete sets of fungi, classified as edible, harmless, poisonous; microscopic fungi hurtful to plants ; insectivorous birds, that all grain or fruit growers should cherish and protect ; sorts of grains or fruits suitable to the various soils of the colony, with specimens of such soils. These are but illustrations of what an Economical Museum should be ; the only difficulty seems to be the drawing a line between the Technological and the Econonic Museums, and I think we ought to deprecate the establishment of mere rival collections. We have not scientists enough to spare men in different places to do the same work, and we have neither the wealth of money or time to spend in running from place to place in our pursuit of knowledge. To the botanist the Botanical Museum, under the care of the Baron von Müeller, offers all that he needs of the flora of Australia ; while our Botanical Garden is not only a thing of beauty, but a live book adorned with nature's own most magnificent paintings, in which those who walk may read and learn.

I have been led thus to take up my time—not intentionally at first—in speaking of the helps we enjoy in this city for the domestication of science. Our one hundred and sixty members show that in this young land minds are not shut to the wonders that nature is ever ready to reveal to those who are willing to open their eyes and see. But with the aid we have—with a land full of unique forms—with many a life history yet unwritten, the worshippers at Nature's altar should be increased many fold, and to all and sundry who wish to do honest work in a humble and patient way our Club holds out a hand of heartiest welcome.

Before I close my address I should note one or two matters of public interest that have engaged the attention of the Club. In October last a deputation waited on the Minister of Lands with reference to the destruction of trees and shrubs in Studley Park. The result has been that increased vigilance has been given to the preservation of the park in its natural beauty. The Club also presented a petition to Parliament supporting the amended Game Act which has abolished swivel and punt guns, by which such wasteful destruction of bird life has been caused for years past on our lakes and swamps. A committee of our Club is also just now busily engaged in preparing a list of insectivorous, game and other birds that, in their opinion, should be brought under the provisions of the Game Act.

As loyal Australians we cannot but be glad that our land is receiving attention from naturalists in the old countries. The miserable description of its fauna and flora given by the elder Darwin, doubtless, as Mr. Lucas pointed out to us at one of our meetings, prevented much attention being paid to our natural history. The advent of Mr. Caldwell, and his patient investigation into the vexed question of the reproduction of the Monotremata and Ceratodus, is a matter for congratulation. Biologists will be eager to learn all he has to tell us. We are glad, also, to welcome to this colony so well known a labourer as Mr. M'Alpin, the newly-appointed lecturer at Ormond College. Congratulating the college on having obtained the services of so distinguished a man, and of one who can make science popular, and is able to lecture on scientific subjects without the continual use of sesquipedal words.

Ladies and gentlemen, while we congratulate ourselves on what has been done, let us ever understand what the true end of science is. It certainly is not the mere gratification of fancies— the passing amusement of an hour. It is not simply to know. Lord Bacon's famous motto was one we might almost take as the motto of our Club—"The true end of science is to enrich human life with useful arts and inventions." And truly, by the patient study of life in all its forms, adding to the sum of human knowledge, we may add to the sum of human happiness. I affirm that that man who helps to make the world cleaner and healthier, or who is able by patient investigation to add to the store of its common wealth, is truly an apostle of that divine kingdom that prophets and seers have forecast and sung of.

In concluding my address, do not suppose that I think for one moment that our young men are ever to be looking through lenses, or that our young women are to do naught but classify flowers or animals. Nor do I suppose or wish that all evening parties should be turned into scientific conversaziones, that lectures should take the place of songs, and dances all give way to disections. I only plead for an intelligent acquaintance with the phenomena of nature, and some knowledge of the laws by which such phenomena are governed ; that the conversation of intelligent people should sometimes rise above the idiotic meanderings of dreary commonplaces, and that blatant ignorance should not assume to be the philosophy of the day. Ladies and gentlemen of our Club, you are doing your part in no unimportant work. You are helping to bring in a time of knowledge that shall alike be useful and reverant. Our land is full of wealth. Rich mines of truth need patient investigation to compel them to yield up their stores—hidden treasures are for these who can learn the password. Let us learn to know that we may know to act.

OUTLINE OF LECTURETTE BY REV. A. W. CRESSWELL, ON SOME OF THE LARGER EXTINCT ANIMALS OF AUSTALASIA.

After a few introductory remarks, the lecturer drew attention to the well-known laws connected with the geographical distribution of animals, according to which every large continental division of the earth has a certain class of existing animals which are more or less peculiar to or characteristic of it, and also the fossil remains of the animals found in the most recent Tertiary deposits of every such " Zoological Province" indicate a pre-existent group of animals of the same types as are now living there, only for the most part on a very gigantic scale. After giving some illustrations of these laws by comparing the Recent with the Pleistocene fauna in the Natural History provinces of Europe, Southern Asia, and South America, the lecturer then proceeded to show that the two provinces of Australia and New Zealand offered no exception. Australia and the adjacent islands formed the great abode of the marsupials, and so also the extinct quadrupeds whose fossil remains were found in the most recent Tertiary formations of Australia were of the same marsupial type, only they were of the most gigantic size, *e.g.*, the Macropus (*Titan* and *Atlas*), and *Procoptodon* (*Goliah*), were the giant prototypes of the kangaroos, only three times as large as the largest " old man." *Diprotodons* (*Australis* and *longiceps*) were the ancient representatives of our little native bear, but were as large as a rhinoceros, and being, of course, unable to climb up trees, used to pull them down, like the *Megatherium*, or giant sloth of South America, and the *Thylacoleos* (*Oweni* and *carnifex*), **or great** marsupial Lions, were the forerunners of the native cats, &c., but were as large as an ordinary lion. They had some interesting peculiarities of dentition, which the lecturer described.

Again, New Zealand was the only present abode of wingless birds of the genus *Apteryx*, or "Ki Wi" of the natives, and that had its great precursor in the *Deinornis* (*giganteus*, *Elephantopus*), or "Moa" of the natives, a bird ranging up to twelve feet high, whose fossil remains are found in the most recent geological deposits of the island, but they had also been found fossil in Queensland.

After referring to some triumphs of Palœontological skill by which some of these creatures had been restored in the first instance from a single tooth or other fragment, and then discovered in more complete form afterwards, so as exactly to "justify the wisdom" of the Palœontologist, the lecturer referred to the "law of correlation of form," and went on to explain from it, how "a single fragment of bone in the hands of a Cuvier, an Owen, or a McCoy, would afford a clue by which any one of these learned savants would be able not only to reconstruct the entire skeleton of the animal to which it belonged, but to predicate its food, its habits, and in a word, its whole natural history." The conclusion of the lecturette consisted of a quotation of Prof. Owen's testimony in favor of the Theistic position as against materialism, as the result of his study of Palœontology.

The lecturette was illustrated by diagrams, and by skulls of recent animals.

The second lecturette, "Insects, their forms and Metamorphoses," was delivered by Mr. F. G. A. Barnard, the hon. secretary, and proved both interesting and instructive. The lecturer, in as simple words as possible, showed the position of the class *Insecta* in the animal kingdom, and its relationship to the other classes of the same sub-kingdom, *Annulosa*. He then gave a brief account of the metamorphosis, or change of form, in the three more or less complete stages through which every insect passes between its birth and its fullest development. The seven principal orders of insects were then rapidly glanced at, and their leading differences explained. References were made to common insects, of the various types, likely to be familiar to most persons, and to a series of drawings made by the lecturer; who, in conclusion, expressed his willingness to afford any information possible respecting the insects in his exhibit in the lower room, as also the *larva* and *pupa* cases of the moths and butterflies shown.

Baron von Mueller, in moving a vote of thanks to the president and lecturers, said, that as one of the earliest naturalists in the colony, it gave him great pleasure to witness the advance and prosperity of the Field Club. A quarter of a century ago, from the chair now occupied by the president, he had prophesied the growth and increased popularity of the study of the Natural Sciences in the colony. He congratulated the president on the use of the happy phrase of the "domestication of science." He was glad to welcome divines like Mr. Halley and Mr. Cresswell amongst the students of science. For the more he and others worked amongst the wonders of Nature, the more impelled they were to recognise a First Great Cause. Mr. A. H. S. Lucas having briefly seconded the resolution, it was carried by acclamation.

The following is a list of the principal exhibitors, and their specimens :—

Mr. D. Best, ten cabinet drawers, containing a fine representative collection of Australian Coleoptera ; and a case of Victorian bird skins.

Mr. F. G. A. Barnard, in illustration of his lecturette, three cases of insects collected in the vicinity of Kew, containing representatives of all the principal orders, sections, families, and genera ; a plan with specimens attached showing the relationship between the different classes of the sub-kingdom *Annulosa;* and also living larva of the Emperor Moth (*Antherœa Eucalypti*), and the pupa cases of several butterflies and moths. Several well-grown Victorian ferns, amongst them being *Gleichenia flabellata, G. circinata, Pteris umbrosa,* and *Schizœa dichotoma.*

Mr. A. J. Campbell, a small case of the nests and eggs of Australian birds ; also the apparatus used for blowing birds' eggs.

Miss F. M. Campbell, a collection of fresh fungi.

Mr. G. Coghill, several pots of Victorian orchids in bloom including *Pterostylis reflexa, P. aphylla, Eriochilus fimbriatus,* &c.

Mr. J. C. Cole, a fine specimen of a fungus growing from the head of the larva of a moth.

Mr. J. E. Dixon, four cases of Victorian fossils, from the Pliocene, Miocene, Eocene, and Silurian formations.

Mr. C. French, a case of Exotic lepidoptera, including the Atlas Moth of China, and other rare species; collection of Goliath beetles from West Africa; a fine pair of living Fijian parrots; also the gold medal and diploma awarded to him for his entomological collection at Amsterdam.

Master C. French, four cases of Victorian and other fossils, &c.; native weapons and utensils from Fiji, New Guinea, and West Australia.

Master G. French, a unique case of Australian and other seeds.

Mr. J. H. Gatliff, five cases of marine shells, comprising 210 species of the genera *Conus, Murex, Voluta, Cymba,* and *Melo.*

Master W. H. F. Hill, two cases of Victorian lepidoptera, result of first and second years' collecting.

Master G. E. F. Hill, two cases of Victorian lepidoptera, result of first and second years' collecting.

Mr. E. E. Johnson, a pelican, and other Victorian birds, cat bird, from Richmond River, red and white coral from Fiji, &c.

Mr. H. Kennon, case containing Victorian and South Sea Island shells, coral, weapons, &c.

Mr. W. Kershaw, two cases Exotic lepidoptera, and two cases of Australian timber-feeding moths.

Mr. T. A. Forbes-Leith, case containing collection of sixty-five Australian parrots; cases containing native cat and kittens; White Goshawks, (male from Gippsland, female from Tasmania); case with opossum mice.

Mr. D. LeSoüef, two live tiger snakes, (venomous), and one live carpet snake, (non-venomous), Victoria; one live diamond snake, (non-venomous), New South Wales; four live Victorian lizards, (blue-tongued, and stump-tailed), and the rare tuatara lizard, (live), from New Zealand; collection of snakes, (in spirits), from Malay Peninsula; a king penguin from Macquarie Island; a cuscus from New Guinea, a small falcon from Malay States, and *Strix scops*, the smallest of the owls from S. Europe, &c.

Mr. A. H. S. Lucas, collection of Victorian sponges.

Dr. Lucas, several rare Victorian moths.

Baron F. von Mueller, wax model of Murray River Lily, (*Crinum flaccidum*), prepared by Mrs Timbrell: specimens in paper of (1), *Rhododendron Toverenæ*, a new species with very large white flower bunches, discovered in New Guinea by Mr. Hunstein, (with woodcut); (2), *Bikkia Bridgeana*, a splendid new species brought from New Guinea by Captain Bridge, R.N.; (3), *Dipteranthemum Crosslandi*, a charming new everlasting, gathered in West Australia by Mr. Crossland, the flower resembling some dipterous insects ; (4), other new plants described in the "Victorian Naturalist;" Edible fruits from New Guinea, viz., *Bassia Erskineana*, *B. Maclayana*, *B. coco*, *Pangium edule*, obtained by Mr. Mikluko-Macklay, and Rev. W. Gill; leaves and acorns of New Guinea Oaks, viz., *Quercus Dalbertisii* and *Q. Gulliveri* ; large Mexican acorns of *Q. Skinneri*; also bound copy of "Eucalyptographia," and plates of forthcoming "Monograph of the Myoporinæ."

Mr. F. Reader, two books of minute Victorian flowering plants, collection of Victorian lichens, including two new species, *Lecanora leucaspida*, Knight, and *Pertusaria albescens*, Knight.

Mr. G. Rose, a case of fossils and minerals.

Mr. A. Thie, a large collection of Fijian and other implements, weapons, manufactures, &c.

Mr. T. Worcester, two cases of land shells, containing many rare species.

About half-past ten the visitors began to disperse, after having spent a very enjoyable and instructive evening.

EXCHANGE.

F. R. would be glad to exchange New Zealand shells, (two glass cases), and many duplicates, named sponges, &c., and Victorian insects and shells, for Australian plants, or books relating to A. botany.

Field Naturalists' Club of Victoria.

OFFICE-BEARERS 1885-86.

President:
REV. J. J. HALLEY.

Vice-Presidents:
MR. T. A. FORBES-LEITH | MR. A. H. S. LUCAS, M.A.

Hon. Treasurer:
MR. E. BAGE.

Hon. Librarian:
MR. C. FRENCH.

Hon. Secretary:
MR. F. G. A. BARNARD, KEW.

Asst. Hon. Secretary:
MR. G. COGHILL.

Committee:
MRS. DOBSON.
„ J. SIMSON.
MR. D. BEST.
MR. J. H. GATLIFF.
„ G. R HILL.
„ D. LeSOUËF.
MR. C. A. TOPP, M.A.

THIS CLUB was founded in 1880 for the purpose of affording observers and lovers of Natural History regular and frequent opportunities for discussing those special subjects in which they are mutually interested; for the Exhibition of Specimens; and for promoting Observations in the Field by means of Excursions to various collecting grounds around the Metropolis.

No Entrance Fee. Annual Subscription, including copy of proceedings 15s., dating from May 1st.

The Ordinary Meetings for the reading of papers, and exhibition of specimens, with a short conversazione are held on the second Monday in each month at the Royal Society's Hall, Victoria Street, Melbourne, at 8 p.m.

The proceedings of the Club are recorded in its journal—the "Victorian Naturalist." Annual Subscription, 6s. 6d.. post free (to members free). The first Volume, comprising sixteen numbers, with title page and index, just completed. Price—Eight shillings (post free).

THE
Victorian Naturalist:

Vol II., No. 3.   JULY 1885.   No 19.

THE FIELD NATURALISTS' CLUB OF VICTORIA.

The monthly meeting of the Club was held at the Royal Society's Hall, on Wednesday evening, 10th June, 1885.

The president, the Rev. J. J. Halley, occupied the chair, and about eighty members and visitors were present.

Among the visitors was Dr. J. E. Taylor, F.G.S., who was received most cordially, and on being introduced to the meeting by the president, briefly thanked the members for the reception given him, and in the course of a few remarks on the characteristics of the Australian fauna and flora, pointed out the peculiar opportunities Australian naturalists' had for finding "missing links"

Correspondence was read from Mr. I. Batey, Sunbury; Mr. S. S. Crispo, Dromana; and others, mainly in support of the Club's proposal *re* Protection of Native Birds.

The hon. librarian reported the receipt of the following additions to the library:—" Proceedings of the Linnean Society, New South Wales," Volume X. Part I., from the society; " Journal of the New York Microscopical Society" Nos. 2 and 3, from the society; " Midland Naturalist," Vols. 1, 2, 3 and 4, from Mr. W. M. Bale; " Proceedings of Ornithological Society of Vienna," from the society.

The hon. secretary reported that the excursion to Lilydale, on the Queen's Birthday, 25th May, was well attended, about twenty-five members and friends being present. Specimens in several departments were rather scarce, but fungi were most abundant, about 100 species being noted. Several good fossils were also obtained at the lime-stones quarries.

The following were elected members of the Club:—Miss Glenross, Mrs. Gunst, Messrs F. E. Hill, S. Lamble, A. Miller, Chas. Officer, jun., William Officer, J. D. Pinnock, D. Strong, and Robert Watson.

The general business consisted of the consideration of Mr. A. J. Campbell's motion in favor of the protection of native birds. A reply was read from the Zoological Society, which recommended that the bee-eaters, wood-swallows, Banksian and Gang Gang Cockatoos should also be included in the list of protected birds. Dr. Dobson thought the list was too long, and that the club would be more likely to be successful if the number of birds to be protected were curtailed, and on his amendment being carried, the list was again referred to the sub-committee, with a view of getting it shortened.

Papers read—1. Mr. A. H. S. Lucas, M.A., read the second part of the paper by Mr. J. B. Gregory and himself, on "An Overland Trip to Wilson's Promontory," giving an interesting account of the natural history of the granitic or southern portion of the promontory. He stated that the locality is well worthy of a visit by students of geology, and of nearly every branch of biology, and in the course of a few years would doubtless become a favorite spot with tourists. He characterized the promontory as the Cornwall of Victoria.

2. Mr. C. French, F.L.S., contributed the seventh part of his paper on "The Orchids of Victoria," in which he described the following species:—*Microtis porrifolia, M. parviflora, M. atraea Corysanthes pruinosa, Pterostylis cucullata, P. furcata, P. reflexa, P. præcox, P. curta, P. nutans;* dried specimens of each of which he exhibited. A short discussion ensued in which Dr. Taylor spoke on the irritability of the labellum in the genus *Pterostylis*, as concerned in the fertilisation of the plants by insects. Mr. C. A. Topp, and Mr. F. G. A. Barnard remarked that though they had carefully observed these flowers, they had never yet noticed any insects about them.

Natural history notes—Mr. C. French, F.L.S., contributed a few remarks and exhibited specimens of the larvæ, etc., of a lepidopterous insect, allied to *Tortricina*, which has recently done much damage amongst cabbage and cauliflower plants. Mr. D. McAlpine spoke of the great importance of studying the economic entomology of the colony.

The following were the principal exhibits:—By Mr. E. Bage, colored plates illustrating "Select Flowers and Fruits of Java" by Madame van Nooten; by Mr. F. G. A. Barnard, Victorian coleoptera, living ferns *Gleichenia circinata*, and *Schizæa dichotoma*; by Miss Campbell, fern new to Victoria, *Polypodium phymatodes*, from East Gippsland, eight dried Victorian ferns including *Botrychium ternatum, B. lunaria, Asplenium nidus, A. flaccidum, Polypodium phymatodes*, and *P. serpens*, new lichens *Usnea retipora* (Knight) Victoria, and *Parmelia Campbellii*, (Knight) New South Wales, also rough drawings of fungi obtained during Lilydale excursion; by Mr. A. J. Campbell, twenty species of rare Australian bird eggs;

by Mr. G. Coghill, orchids in bloom, *Pterostylis præcox, P. nutans* and *P. concinna*; by Mr. A. Coles a very fine *Ornithorhynchus* twenty-three inches long, also Victorian game birds; by Rev. A. W. Cresswell, fossils from Lilydale; by Mr. J. E. Dixon, older pliocene fossils from Cheltenham; by Mr. C. French F L.S., exotic coleoptera, family *Cetoniidæ*, orchids in illustration of paper, and cabbage moth in various stages; by Master C. French, fossils from Cheltenham; by Mr. J. H. Gatliff, Victorian shells, eighteen species of family *Veneridæ*; viz.. *Rupellaria* (3 *sp.*), *Tapes* (1), *Venus* (6), *Cytherea* (4), *Meroe* (1), and *Dosinia* (2); by Rev. J. J. Halley, specimens of limestone from the Great Pyramid; by Miss Halley, nests of weaver bird, India; by Masters Hill, Victorian lepidoptera; by Mr. H. W. Hunt, Victorian birds' eggs; by Mr. H. Kennon sea-gulls (living) from Warrnambool; by Mr. W. Kershaw, Australian lepidoptera; by Mr. T. A. Forbes-Leith, eighty-four Victorian birds representing thirty families; by Mr. A. H. S. Lucas, M.A., Victorian *Asteridæ*, and plants and shells from Wilson's Promontory in illustration of paper; by Mr. D. McAlpine, frog in first stage of new process of dry preservation ; by Mr. F. Reader, plants from Studley Park, (*Coniferæ* to *Amaryllidæ*); by Mrs J. Simson, painting on cobweb from Innsprück, and picture in colored sand from Isle of Wight; by Miss Mary Simson, a flying mouse from Langi Kal Kal, Victoria; by Mr. A. Thie, English Ammonites; by Mr. H. Watts, a number of objects under the microscope.

After the usual *coversazione* the meeting terminated.

EXCURSION OF THE FIELD NATURALISTS' CLUB.

THE last excursion prior to the annual meeting of this Club took place on Saturday, May 9th, under the leadership of Mr. C. French, the locality chosen being as stated in the annual report, Brighton, because of its convenience and correspondingly superior resources, which in the short autumnal days and the little time at the disposal of members leaving by the 2 p.m. train, is of great consideration. Briefly then, the members left Melbourne by the 2 p.m. train, and on its arrival at Brighton, it was found that notwithstanding the threatening appearance of the weather, there was a fair attendance, including two lady members of the Club. To save time in walking, a conveyance was in readiness, and drove the party to within a few hundred yards of the Red Bluff Hotel, when a start was made inland. Plants in bloom were but few, although we soon came across a patch of damp, heathy country, in which grew quantities of the plants usually to be found in the district. *Pterostylis aphylla*, and

Eriochilus fimbriatus, were here in considerable numbers, and were secured for the purpose of either cultivation or herbaria. Proceeding onwards towards the hills, we find our old favorite *Styphelia humifusa*, in full bloom, its beautiful crimson tubular-shaped flowers rendering it a general favorite. Several specimens of the *Prasophyllum* found on last trip were also seen, and as this small species may not be either *P. archeri*, or *P. intricatum*, a sharp look-out in the early part of April next, should be kept. As we steer towards the flat or swamp known to old colonists as the Hawk's nest, we pass numerous specimens in flower of *Monotoca scoparia*, and *Epacris impressa*, which by the way reminds us that it was only about two miles from this spot, where was found the first specimen known of the beautiful "double white" variety of this species, and which now is, or was, in the collection of Mr. Scott, of the Royal Nursery, Hawthorn. Birds hereabouts are few, although we were informed that there were quail to be found not far from us, and a few specimens of the Honey-eaters, *Meliphagidæ*, some little Wrens, and a bronze-winged Pigeon, were about the only "land birds" seen. As the afternoon wore on, and we were anxious to do what we could, after collecting a few specimens of the common but very curious lichen, *Cladonia retipora*, which was in fine fruit, and three specimens of *Pterostylis nana*, (the only ones seen during the trip,) we steer a bee-line for the coast, collecting on our way bouquets of wild flowers, as *Epacris*, *Styphelia*, *Banksia*, *Acacia suaveolens*, which together with the curious bronzy-coloured *Restiaceous Hypolæna fastigiata*, made a very pretty bunch, in which the *Correas* and *Hibbertias* were prominent. The belt of scrub being reached, a search was made for the early orchids, and very soon was found *Pterostylis vittata*, and *Acianthus exsertus*. The *Jungermanniœ* were seen, but as it is too early for fruiting specimens, they were passed, or left for a future occasion. To those interested in spiders, it may be mentioned that a very fair collection might be made in the district, and some of these species are very handsome. As it was now getting dusk a start was made for the Red Bluff Hotel, and although barely able to distinguish one plant from another, several pretty mosses and huge *Polypori* were seen, in company with numerous other fungi. A specimen of *Lasiopetalum Baueri*, now somewhat rare about Melbourne, was here found growing just above high water mark, near which were specimens of *Lobelia anceps*, *Salicornia*, *Aster*, and other sea coast plants. On reaching the hotel, the conveyance being in waiting, the railway station was reached in good time, and thus a very pleasant afternoon had been passed. It is to be hoped that during the present year of the Club's existence, these excursions will be better attended as much may be gained thereby physically as well as intellectually. Melbourne was reached at about seven o'clock.

THE QUEEN'S BIRTHDAY EXCURSION. TO LILYDALE.

TAKING advantage of the holiday on Monday, 25th May, a Club excursion was arranged for that day, and after some little discussion at the previous monthly meeting, it was decided to visit Lilydale, as being perhaps the most promising locality at that season of the year. Accordingly at 6.15 a.m., about sixteen members of the Club, including two or three ladies, met at the Prince's Bridge station, and leaving town a few minutes after, in due course arrived at Lilydale. Several more members were picked up at the suburban stations, and at Lilydale three others appeared, who had gone up on the Saturday and Sunday, making altogether a party of about twenty-five. On arrival at Lilydale, it was decided to explore the valley of the Olinda Creek as being the most likely direction to reward the trouble of such early rising. Two parties were now formed, those intent on shooting going on first, the arrangement being made to unite again at a pretty spot on the creek known to the leaders. The rest of the party, consisting principally of botanists and entomologists, after despatching a late breakfast at Lithgow's, started off towards the creek, keeping in a south-easterly direction, and were soon in scrubby country. The entomologists turned over logs and stripped the bark off trees in vain, nothing of any interest being obtained; flowering plants were also very scarce, but fungi were most abundant, and our mycologist had almost as much as she could do in collecting specimens, or packing away those brought to her by other members of the party. On the way several species of ferns were secured by those wanting them, a fine patch of *Gleichenia circinata*, being quite despoiled of its starry fronds for making "bird's nests." The sportsmen were now met, but with almost empty bags. For some unaccountable reason, the valley, usually a good collecting ground, was almost devoid of bird life. The only birds seen during the trip being the frontal shrike-tit (*Falcunculus frontatus*), yellow robin (*Eopsaltria Australis*), Tasmanian honey-eater (*Meliornis Australasiana*), spine-billed honey-eater (*Acanthorhynchus tenuirostris*), gang-gang cockatoo (*Callocephalon galeatum*), and Pennant's parrakeet (*Platycercus Pennantii*.)

A cutting at a bend of the creek was now reached, the damp sides of which were clothed with young seedling ferns of different varieties, in many stages of growth, also several beautiful species of fungi. A short distance a-head the camp fire was seen, and on reaching the spot a halt was made, and the luncheon baskets relieved of some of their good things. One of the members having offered his services as cook, tea was soon manufactured in the orthodox Australian style, and a vegetable beef-steak (*Fistulina hepatica*) cooked. However this latter proved uneatable, being too old. Near here some splendid specimens of the larger star-fern (*G. flabellata*) were obtained, and a log over the creek was found covered with the delicate little *Hymenophyllum Tunbridgense*. About fifteen other species of ferns were seen during the excursion, but none of them were particularly rare.

After luncheon, the party guided by two members who had been over the same ground on the previous day, rambled towards the ranges, on the way obtaining the only orchid found in bloom, *Pterostylis parviflora*. A little further on the foot-hills were reached. Here the native heath (*Epacris impressa*,) principally the pink and crimson varieties, was found in great abundance, and large bunches were gathered for home decoration. A little higher up *Grevillea sp.* was found in flower. The road leads up on to the top of the range affording several pretty views on the way, and crossing one or two fern gullies. At the back of the range many splendid fern gullies exist, which would repay a search at some future time. In one of these visited on the previous day by the members before referred to, several small specimens of the pretty fern *Osmunda barbara* were obtained, and a delicate little blue fungus *Agaricus* (*Leptoma*) *sp.* A return was now made, a single specimen of *Comesperma ericinum*, being noted in flower. Another short halt was made at the creek, after which a different route to that of the morning was taken over the hills towards Lilydale. On the way several plants of orchids, probably *Pterostylis curta* were seen, also some large fungi, *Agaricus sp.*, which were pronounced edible by our mycologist. After a pleasant ramble Lilydale was reached in ample time to allow of a short stoppage for open air tea, before wending our way to the train.

As fungi were the most numerous of the specimens collected, a more detailed list of the species by Miss Campbell may prove interesting.

There were over thirty-five Agarics found, these include many edible kinds as well as the common mushroom, the white lady, and the beautiful *Cantharellus*; gay coloured ones, as the bright red, yellow, and green *Russulas*, the luminous *Panus incandescens*, the dainty little gray Agaric, smelling as sweet and strong as violets, the delicate *Xerotus*, the fast fading *Caprinus*, tiny exquisite blue Agaric (*Leptoma*) of Mr. Tisdall's paper, some whose acrid taste, gave warning of poison, and a large bright violet-coloured Agaric which is edible; the three *Boleti* did not look or smell so tempting as usual. About ten *Polypori*, of these *P. Cinnabarinus* attracting most attention, with its bright red colour; the specimens of *Fistulina hepatica*, the celebrated beef-steak fungus were too old to be eaten; a purple *Trametes*; two *Hydnums* one jelly-like, pale lavender spines, very good eating; three *Storeums*, one for its elegant form and pretty markings continually picked up; the small bright yellow, jelly-like *Guepinia Spattularia*, was plentiful; many species of *Clavaria* were plentiful, *C. botrytis, C. aurea*, etc.; a white jelly-like *Tremella* which can be eaten when fresh; the net puff ball, *Tleodictyon gracile* which is eaten by the New Zealanders; common puff ball, *Lycoperdon gemmatum*; two bright red and a yellow cup-like *Peziza*; *Leotia lubrica* had the most peculiar appearance of any found, bright yellow, semi-transparent stipe, brownish yellow pileus; an uncommon *Hypoxylon*; also very many micro-fungi brought the number up close to a hundred distinct species for the day.

The two members out on the previous day, found a fine patch of the ordinary mushroom, *Agaricus campestris*, one of which was quite six inches in diameter, growing on a roadside.

The geologist of the party, the Rev. A. W. Cresswell, who spent the day at the Cave Hill limestone quarries, furnishes the following account of his experiences:—

Went to the limestone quarries, about half a mile S.E. of Lilydale; being only in search of fossils this time, did not make accurate stratigraphical observations. The quarry, however, is well known to be in a limestone, granular, crystalline, marble formation, about 100 feet thick, interstratified with the upper Silurian rocks, (sandstones, schists, &c.,) of the district. It is not thought to be very persistent or to extend any great distance along its strike (which is nearly meridional as usual with the Silurian,) but is believed to be more or less lenticular and to thin out at no great distance north and south. The prevailing colour is cream-coloured, but some of the strata are of a ferruginous red and others of a bluish grey. The dip is about 45 degrees east, but exact statigraphical details will be found in the Geological Survey Report for 1855-56. I had never seen any fossils in it before, except a few *Favosites* (Millipore corals) here and there where the surface is weathered, but this last time succeeded in getting the following fossils, viz.:—Several specimens of a sub-genus of Turbo, one of them being as large as a good sized recent *Turbo undulatus*. The form appears to me to be close to *Euchelus*, there being no umbilicus or the columella showing trace of being toothed, it is very like our common recent *Euchelus canaliculatus*, but has finer and more numerous liræ. The nearest shell to it in Murchison's "Siluria" appears to be *Cyclonema corallii* of the Upper Ludlow, with which it is perhaps identical. Several specimens of *Murchisonia* apparently corresponding to *M. corallii* of the Upper Ludlow as figured in Murchison's "Siluria." A *Bellerophon*, which I do not know the specific name of, and will have to show to Prof. McCoy for exact identification; and several specimens of the common Upper Silurian species of *Favosites* called *Favosites aspera*, (one of which is sent herewith); a single joint of *Crinoid* stem, probably an *Actinocrinus*. A mammillary stalactite from the roof of a cavern leading into the quarry was also obtained. It should be mentioned that microscopic sections of this marble show a partly brecciated and partly oolitic structure.

THE PHANEROGAMOUS PLANTS OF STUDLEY PARK, KEW, NEAR MELBOURNE.

By F. Reader.

Read before the Field Naturalists' Club of Victoria, Feb. 10, 1885.

(Part III.)

Order, Sapindaceæ, A. L. de Jussien.

Properties.—Various. Root of *Cardiospermum Halicacabum* is aperient, *Sapindus Saponaria* yields a detersive and acrid fruit, containing *Saponin*. The tincture of the berries employed in chlorosis. American Acer species yield maple sugar. Guarana, from the seeds of *Paullinia sorbilis*, Mart, the Braz. Cocoa, contains a large amount of Guaranine, an active, bitter principle, said to be identical with Caffeine. Guarana is employed in nervous headache.

Dodonæa, L. Etym. Named in honor of Dodonæus, a celebrated physician and botanist at the time of the emperors, Maximilian II. and Rudolph II.

D. viscosa, L. Syn. D. viscosa, var. vulgaris, Benth. Vern., name, Switch Sorrel. Flowers March. Distributed New Zealand and Tasmania. In New Zealand it is called Akerautangi, ake, and the wood is used for native clubs.

Order, Portulaceæ, A. L. de Jussien.

Properties.—Purslane, Portulaca oleracea L, and others are employed as esculents.

Claytonia, Gronovius. Etym. In honor of Dr. T. Clayton, a Kentish physician and botanist.

C. calyptrata, F. von Mueller. Flowers September to December. Distributed Tasmania.

Order, Caryophylleæ. Scopoli.

Properties.—Unimportant. Silene Virginica is said to have an anthelmintic root. Saponaria officinalis, and Gypsophila Struthium were formerly used as aperients in skin diseases; they contain Saponin. *Lychnis, Githago, Lam.*, the Corn Cockle, now introduced with cereals around Melbourne, also contains Saponin in the seeds and Agrostermmin.

Spergularia, Persoon. Etym. The diminutive of Spergula, from Spargo, alluding to the seeds being widely scattered.

S. rubra, Camb. Vern. name, Sand-wort, Spurrey. Flowers September to January. Distributed. Except the Arctic and Tropic zones extends to nearly all countries.

Stellaria, L. Etym. from stella, indicating the star-shaped spreading of the corolla. Vern. name, Star-wort, Stitch-wort.

S. pungens, Brongn. Flowers September to December. Distributed Tasmania.

S. media, Villars, Chickweed. Flowers nearly all the year round. Distributed, through cultivation dispersed over nearly all temperate and arctic regions. Naturalized.

Uses.—Formerly in repute for Phthisis, dysentery, hæmorrhage and diseases of the skin, &c. Poultry and small birds are fond of the whole plant, especially the seeds.

Cerastium, L. Etym. From the Greek κέρας (keras), alluding to the curved capsules of some species. Vern. name, Mouse-Ear Chickweed.

C. glomeratum, Thuellier. Syn., C. vulgatum, L. Flowers nearly all the year round. An alien. Distributed all temperate and cold regions.

Spergula L. Etym. From spargo, in allusion to the scattering of its seeds. Vern., name, Spurrey.

S. arvensis, L Corn or Field Spurrey. Flowers September to March. An alien. Distributed Arctic Europe, North Africa, West Asia, to North West India. Introduced in North America.

Uses.—Cattle are fond of this plant.

Polycarpon, L. Etym From the Greek, πόλυς (polus) and καρπος (karpos), in allusion to the abundant capsules. Vern. name All seed.

P. tetraphyllum, L. Flowers November to March. Distributed. Almost universally dispersed within the warm and temperate zone.

Silene, L. Etym. From the Greek ςίαλον (sialon), saliva alluding to the viscidity of many species.

S. gallica, L. Flowers September to January. An alien. Distributed. Represented in most cultivated regions. There is a variety the Silene *quinquevulnera, L.*, with petals entire and spotted with red. Frequently growing with *S. gallica.*

S. pendula, L. Flowers October, November. Probably escaped from the gardens. Native of Sicily.

Order Amarantaceæ, A. L de Jussien.

Properties.—Unimportant. On account of their often richly coloured flowers mostly cultivated as pot herbs, &c..

Alternanthera, Forskæl. Etym. Alluding to the fertile stamens alternating usually with antherless filaments,

A. triandra, Lamarck., Syn., A. sessilis, Br. A. denticulata, A. Cunn. Flowers January to June. Distributed Warmer regions of Asia and America, Africa, Tasmania and New Zealand. Variable.

Order, Salsolaceæ, L.

Properties.—Various. *Chenopodium Quinoa* is widely used as an article of food in Peru. *Ch. anthelminticum* yields Wormseed Oil, an effective vermifuge. Spinach, Beet and others are esculents. *Salsola* and *Salicornia* furnish Carbonate of Soda. From Beet-roots, a fine sugar is extensively manufactured. *Chenopodium olidum* and *baryosmon* act as fœtid emmenagogues. *Ch. ambrosioides* is an aromatic expectorant, &c. Many are known as pot herbs.

Rhagodia, R. Br. Etym. From rhax, berry, alluding to the numerous berry-like fruits. Vern. name, Red or Sea-berry.

R. nutans, R. Br. Flowers November to January. Distributed Tasmania.

Chenopodium Tournef. Etym. From the greek, $\chi\bar{\eta}\nu$, $\chi\eta\nu\dot{o}\varsigma$, (chen, chenos,) goose, and $\pi o\bar{v}\varsigma$ (pous) foot; the leaves of some species supposed to resemble in shape the foot of a goose. Vern. name, Goosefoot.

Ch. murale, L. Vern. name, nettle-leaved Goosefoot. Introduced Distributed Europe, North Africa, W. Asia, to N. W. India; introduced in North America, Tasmania and New Zealand.

Ch. album, L. Vern. name, white Goosefoot. Introduced, Distributed. Through colonisation dispersed over all zones. Contains Chenopodin.

Enchylæna, R. Br. Etym. Alluding to the succulent calyx. giving the fruit the appearance of a berry.

E. tomentosa, R. Br. Flowers September. Distributed throughout Australia.

Order Polygonaceæ, A. L. De Jussien.

Properties—Often astringent and purgative; some species yield oxalic and malic acids; the seeds of others are farinaceous and esculent. The all important Rhubarb is the most important in the order. *Coccoloba uvifera, Jacq.*, contains kino an astringent. Some species of *Polygonum* yield Indigo. Many species of *Rumex* contain in the root Chrysophanic acid, employed in Psoriasis, &c.

Polygonum, Tournef. Etym. From the greek $\pi o\lambda\acute{v}\varsigma$ (polus) many, and $\gamma o\nu v$ (gonu) knee, alluding to the many joints of the stem and branches. Vern. name, Bistort or Persicaria.

P. strigosum, R. Br. Flowers December to March. Distributed Tasmania.

P. minus, Hudson. Flowers January to June. Distributed Tasmania and New Zealand.

P. aviculare, L. Vern. name, Knotgrass, Wire Weed, Hogggrass. Flowers nearly all the year round. An alien. Distributed almost cosmopolite.

Uses.—A mild astringent, Diarrhœa, &c., externally for wounds. Fruit emetic and cathartic. The whole plant yields indigo-blue.

P. Hydropiper, L. Vern. name, water pepper or biting Persicaria. Flowers February June. Distributed North Temperate Hemisphere and to Australia.

Uses.—Contains Polygonic acid of an acrid, bitter taste and a volatile acrid principle. Plant when chewed imparts a hot and pungent taste to the tongue.

P. prostratum, R. Br. Flowers March April. Distributed Tasmania and New Zealand.

Muehlenbeckia, Meissner. Etym. In honor of Dr. Muehlenbeck, who closely studied the plants of Alsace.

M. adpressa, Meiss. Vern. name, Sarsaparilla, of the colonists. Flowers September November.

Uses.—Produces the same effect as the true *Smilax* species employed as alteratives and tonics.

Rumex, L. Etym. The old latin name, alluding to some resemblance of the leaves to the Roman war-arms. Vern. name, Dock and Sorrel.

R. crispus, L. Vern. name, Curled Dock. Flowers nearly all the year round. Distributed Europe, North Africa, Temperate Asia to Japan. Introduced in North America, Tasmania and New Zealand.

Uses.—Is an alterative, detergent and antiscorbutic remedy, mildly aperient, acting on the colon ; may be given in Scrofula, cutaneous eruptions, and in the form of an ointment. Leaves may be advantageously used as an article of diet in scurvy, in the form of a salad. They are agreeably acid to the taste, owing to binoxalate of potash with tartaric acid, but lose their taste in drying. Juice of the leaves mixed with water affords an agreeable acidulous drink. The cortical part of the root is the most active. In America the concentrated tincture of Rumex is now prepared and used for the complaints above mentioned.

R. Acetosella, L. Vern. name, Sheep's Sorrel. Flowers nearly all the year round. Distributed. Widely diffused throughout Temperate and Arctic zones.

Uses.—Leaves used as a salad; abound in binoxalate of potash. 100lb. of the leaves yield 8lb. of the salt. The decoction of the root or the powdered root are refrigerants and effectual anthelmintics. The seeds are astringent and useful in Hœmorrhage In arid ground and at the end of summer the whole plant assumes a bright red colour,

R. Brownii, Campdera. Flowers November to May. Distributed Tasmania.

R. bidens, R. Br. Flowers December to March. Distributed Tasmania.

ERRATA.

On page 27 of Vol. II., No. 2, line 19, for " Flowers September to January," read " nearly all the year round."

On same page after. " *Casuarina, Rumphius. Etym.*," read : " Supposed to allude to the leaves resembling the feathers of the Cassowary."

NOTES.

MICRO-FUNGI.

At the last meeting of the Microscopical Society, one of the Vice-Presidents, Mr. F. Barnard, of Kew, also a member of the F.N. Club, read some notes on Micro-fungi recently forwarded to England by him for naming. Several of these have proved to be new, and one *Phragmidium Barnardi* (Plow.), found on a species of *Rubus* in Studley Park, has been named after him. We understand Mr. Barnard will be glad to receive specimens of Micro-fungi from our country friends, in exchange for other microscopic objects.

THE NATIONAL MUSEUM.

It will be remembered that in February last a deputation waited upon the Trustees of the National Museum, with reference to affording greater facilities to students at that institution, and were promised that the Director, Professor McCoy, should report upon the suggestions then made. His report, which is too long for publication here, appeared in the *Argus* of June 1st, but as it seemed the ideas of the deputation had been somewhat misunderstood, the Committee of the Club felt themselves called upon to make a further representation of the matter, and have forwarded a letter to the Trustees in reply, which will be found at length in the *Argus* of June 19th.

EXCHANGES.

H. Watts would exchange Sea-weeds, mounted and named, for Australian Ferns, and would be glad to correspond with any person who could obtain parasites of Birds and Animals, either for exchange or otherwise. Address, 20 Wellington Street, Collingwood.

F. R. would be glad to exchange European Micro-Fungi for Australian Plants, or back numbers of "Southern Science Record." 46 Victoria Street, West Melbourne.

REIMS
CHAMPAGNE

Has the Largest Sale of any Champagne.

SOLE AGENTS FOR VICTORIA—

ALEX. JOSKE & CO.,

16 Little Collins Street East.

WM. S. HUSBANDS,
Manufacturing Optician
81 QUEEN STREET,

MELBOURNE,

(Established 1862),

BRISTOL, ENGLAND, (Established 1762.)

Students Compound Achromatic Microscopes and Accessories in Stock

Mathematical, Philosophical, Meterological, Nautical Optical, Surveying and Mining Instruments, Importer, &c.

The Metfords Theodolites and Level Combined
The Improved Dumping Levels Ball Motion

Vol. II. No. 2.　　　　　　　　　　　　　June 1885.

THE

Victorian Naturalist:

THE JOURNAL AND MAGAZINE

OF THE

Field Naturalists' Club of Victoria.

The Author of each article is responsible for the facts and opinions he records.

CONTENTS:

	PAGE
Proceedings of the Field Naturalists' Club of Victoria ...	17
Succinct Notes on some Plants from New Guinea. By Baron Ferd. von Mueller, K.C.M.G.	18
Charles Darwin on Australia. By A. H. S. Lucas, M.A.	20
The Phanerogamous Plants of Studley Park, Kew, near Melbourne. By F. Reader.	24
Notes.	28

PRICE—SIXPENCE

Emerald Hill:
J. C. MITCHELL, PRINTER, CLARENDON ST.
1885.

CPSIA information can be obtained
at www.ICGtesting.com
Printed in the USA
BVHW061459071118
532207BV00059B/3215/P